SpringerBriefs in Molecular Science

Green Chemistry for Sustainability

W0044173

Series editor

Sanjay K. Sharma, Jaipur, India

More information about this series at http://www.springer.com/series/10045

Mohamed A. Barakat · Rajeev Kumar

Photocatalytic Activity Enhancement of Titanium Dioxide Nanoparticles

Degradation of Pollutants in Wastewater

 Springer

Mohamed A. Barakat
Department of Environmental Sciences
Faculty of Meteorology and Environment
King Abdulaziz University (KAU)
Jeddah
Saudi Arabia

and

Central Metallurgical R&D Institute
Helwan, Cairo
Egypt

Rajeev Kumar
Department of Environmental Sciences
Faculty of Meteorology and Environment
King Abdulaziz University (KAU)
Jeddah
Saudi Arabia

ISSN 2191-5407   ISSN 2191-5415   (electronic)
SpringerBriefs in Molecular Science
ISSN 2212-9898
SpringerBriefs in Green Chemistry for Sustainability
ISBN 978-3-319-24269-9   ISBN 978-3-319-24271-2   (eBook)
DOI 10.1007/978-3-319-24271-2

Library of Congress Control Number: 2015948850

Springer Cham Heidelberg New York Dordrecht London

Printed on acid-free paper

Springer International Publishing AG Switzerland is part of Springer Science+Business Media
(www.springer.com)

Contents

Abstract

Advanced oxidation processes (AOPs) with UV irradiation and photocatalyst titanium dioxide (TiO_2) are gaining growing acceptance as an effective wastewater treatment method. A comprehensive study of the UV/visible-TiO_2 photocatalytic oxidation process was conducted with an insight into the mechanism involved, catalyst TiO_2, irradiation sources, types of reactors, comparison between effective modes of TiO_2 application as immobilized on surface or as suspension. Photocatalytic degradation technique with TiO_2 is generally applied for treating wastewater containing organic contaminants due to its ability to achieve complete mineralization of the organic contaminants under mild conditions such as ambient temperature and ambient pressure. TiO_2 is highly stable in aqueous media and is tolerant to both acidic and alkaline solutions. It is inexpensive, recyclable, reusable, and relatively simple to produce. The large band gap of TiO_2 lies in the UV range, which allows for only 5–8 % of sunlight to be useful for the activation of the catalyst. Therefore, a visible light activated catalyst is desired that can take advantage of a larger fraction of the solar spectrum and would be much more effective in environmental cleanup. Several competing effects inherently limit catalyst efficiency. The positively charged holes and negatively charged electrons tend, by nature, to recombine to yield a neutral state, through emission of photons or phonons. This can occur via volumetric or surface recombination.

There are several known ways to increase the efficiency of a photocatalyst. The effects of particle size reduction, anion doping with nitrogen ions, and cations, e.g., Nd, Pd, Pt, Co doping have been extensively researched in order to increase its effectiveness by introducing trapping sites. The aim of this chapter is to give an overview of the development and enhancement of the activity of TiO_2 nanoparticles in photocatalysis. The topics covered include a detailed look at the unique properties of the TiO_2 nanoparticles and their relation to photocatalytic properties. The utilization of the TiO_2 nanoparticles as photocatalysts, in the non doped and doped forms, has been reviewed. Finally, the use of modified TiO_2 nanoparticles has made a significant contribution in providing definitive mechanistic information regarding the visible light photocatalytic process.

Keywords AOPs · Photocatalysis enhancement · Photocatalysis doping · Photocatalysis wastewater · TiO_2 nanomaterials water pollution · Wastewater treatment nanoparticles

Introduction

Oxidation technologies for the wastewater treatment are widely investigated for the degradation of organic contaminants. Ozonation, H_2O_2 based methods, photochemical, wet oxidation electrochemical, and photocatalysis, etc. have been used for this purpose [1]. Heterogeneous photocatalysis is one of the most advanced oxidation process applicable to decontamination of water. According to the IUPAC, [2] a photocatalyst is a solid material that is able to produce, by absorption of light quanta, chemical transformations of the reaction participants, repeatedly coming with them into the intermediate chemical interactions and regenerating its chemical composition after each cycle of such interactions. Accordingly, photocatalysis is defined as a "change in the rate of a chemical reaction or its initiation under the action of light (photons) in the presence of a substance—the photocatalyst—that absorbs light and is involved in the chemical transformation of the reaction partners" [1, 3]. The major advantages of photocatalysis in wastewater treatments are:

(i) Photocatalysis using visible light offers a good substitute for the energy-driven wastewater treatment methods.
(ii) Photocatalysis can be used for the degradation or change a variety of harmful components to low toxic or non-toxic products and generation of secondary waste is minimal.
(iii) This process can be applied to aqueous and gaseous-phase treatments, as well as solid- (soil-) phase treatments to some extent.
(iv) The reaction conditions for photocatalysis are mild, the reaction time is modest, and a lesser chemical input is required.

(v) The option for recovery can also be explored for metals, which are converted to their less-toxic/nontoxic metallic states.

(vii) The photocatalyst remains chemically unchanged during and after photocatalysis and can be reused several times.

References

1. M. Luisa Marin, Lucas Santos-Juanes, Antonio Arques, Ana M. Amat, and Miguel A. Miranda. Organic Photocatalysts for the Oxidation of Pollutants and Model Compounds. Chem. Rev. 2012, 112, 1710–1750.
2. Braslavsky, S.E. Glossary of terms used in photochemistry, 3rd edition (IUPAC Recommendations 2006). Pure Appl. Chem. 2007, 79, 29–465.
3. B. Ohtani Principle of Photocatalysisand Design of Active Photocatalysts Chapter 5 in New and Future Developments in Catalysis. 2013, Pages 121–144.

Photocatalytic Activity Enhancement of Titanium Dioxide Nanoparticles

1 Fundamentals and Mechanism of TiO$_2$ Photocatalysis

Photocatalysis is based on the excitation of a semiconductor material with irradiation of light energy at least equal or greater to that of the band gap energy of the material. An electron in an electron filled valence band (VB) is excited upon irradiation of light to a vacant conduction band (CB) and leave behind a positive hole (h$^+$) in the VB. These electrons and holes (e$^-$ and h$^+$) are mainly responsible for the generation of active species which degrade the target molecules. These charge carriers e$^-$ and h$^+$ drive the reduction and oxidation, respectively [1, 3, 4]. These charge carriers migrate to the photocatalyst surface and indicate the secondary reactions with the adsorbed components on the surface of solid semiconductor. The photo excited e$^-$ in CB changes oxygen to the superoxide radicals and hydroperoxide radicals while positive h$^+$ in VB can oxidize the adsorbed water or hydroxyl ions to from hydroxyl radicals (Fig. 1). These reactive species (superoxide radicals or hydroperoxide radicals and hydroxyl radicals) can be involved in the degradation of surface adsorbed compounds. The overall process can be summarizing as follows: Semiconductor material

© The Author(s) 2016
M.A. Barakat and R. Kumar, *Photocatalytic Activity Enhancement of Titanium Dioxide Nanoparticles*, SpringerBriefs in Green Chemistry for Sustainability, DOI 10.1007/978-3-319-24271-2_1

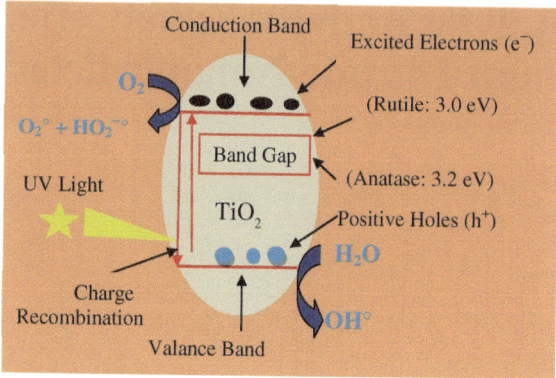

Fig. 1 UV light TiO_2 photocatalysis mechanism

$$(TiO_2) + h\nu \rightarrow h_{VB}^+ + e_{CB}^- \tag{1}$$

$$e_{CB}^- + O_2 \rightarrow {}^\bullet O_2 \; (\text{reduction}) \tag{2}$$

$$h_{VB}^+ + H_2O \rightarrow {}^\bullet OH + H^+ \, (\text{oxidation}) \tag{3}$$

$${}^\bullet O_2 + H^+ \rightarrow HO_2{}^\bullet \tag{4}$$

$$e_{CB}^- + HO_2{}^\bullet \rightarrow HO_2^- \tag{5}$$

$$HO_2^- + h_{VB}^+ \rightarrow H_2O_2 \tag{6}$$

Titanium dioxide is one of the most investigated nanomaterial during last decades with the major focus in the area of energy and environmental applications. The major breakthrough come in 1972 when Fujishima and Honda discovered the photosensitization effect of TiO_2 electrode for the electrolysis of water into hydrogen and oxygen in UV light irradiation [5]. Since 1972, a lot of photocatalytic studies have been reported on the TiO_2 based catalyst because of its high surface area, biological and chemical stability, low cost, low toxicity and high photocatalytic activity. Generally, TiO_2 existed in four different phases, anatase (tetragonal), rutile (tetragonal), brookite (orthorhombic) and TiO_2 (B) (monoclinic). TiO_2 in anatase form shows the highly photocatalytic activity compared to rutile and brookite [6, 7]. All structures are built-up of TiO_6 octahedra but differ

Table 1 Crystallographic data of different TiO$_2$ polymorphs [7]

	Rutile	Anatase	Brookite
Crystallographic structure	Tetragonal	Tetragoanl	Orthorhombic
Space group	P4$_2$/mnm	I4$_1$/amd	Pcab
Cell parameters (A)	a = 4,5933	a = 3,7852	a = 9,1819
	b = –	b = –	b = 4,4558
	c = 2,9592	c = 9,5139	c = 5,1429
Z (molecules/cell)	2	4	8
Structure			
Ti coordinance	6	6	6
Density	4,24	3,83	4,17
Ti–O distances (A)	2 at 1,946	2 at 1,937	2 at 1,993
	4 at 1,984	4 at 1,964	1 at 1,865
			1 at 1,919
			1 at 1,945
			1 at 2,040

in their stacking. The structures of all three forms are shown in Table 1 [7].

TiO$_2$ photocatalyst is only restrict to absorb the light in ultraviolet region (wavelength < 390 nm) due to the wide band gap (3.2 eV for anatase and 3.0 eV for rutile). The photoexcitation of TiO$_2$ generate the electon/hole pairs. The UV light activity and fast recombination of electon/hole pairs limits the use of TiO$_2$ in visible light and the photocatalytic efficacy, respectively.

TiO$_2$ photocatalyst is only restrict to absorb the light in ultraviolet region (wavelength < 390 nm) due to the wide band gap (3.2 eV for anatase and 3.0 eV for rutile). Although, solar light contains only about 2–3 % UV light, therefore, TiO$_2$ based materials which can harvest solar light are always in need. Modifications by various strategies have been applied to make the TiO$_2$ as visible light active material such as doping with metal and non metals, co-doping, coupling etc. [6, 7].

2 Modification of TiO$_2$

In last decade, new modified TiO$_2$ based photocatalyst for the visible light photocatalytic activity have been developed by tailoring its surface by using several approach. Doping of impurities, coupling with narrow band gap semiconductor for shifting TiO$_2$ response to visible region are the most promising approaches. It is desirable to sustain the integrity of the crystal structure of the photocatalyst while changing its electronic structure [8]. There are several approaches which can be used to modify and enhance the visible light photo-catalytic activity of TiO$_2$ such as doping, co-doping, coupling/composite, hybridization, doping and coupling, capping or coating etc.

Modification of the TiO$_2$ not only reduces its energy band gap but also prevent electron/hole pair charge recombination which enhanced the photocatalytic activity.

3 Doping

Doping of metal or nonmetal has been widely used to tune the optical band gap of the TiO$_2$ for the catalytic purpose. The doping process is an effective approach to improve the visible light activity. Doping with cations or anions may alter the band gap or introduction of intra band gap states which may make TiO$_2$ to absorb the visible light.

4 Non Metal Doping and Co-Doping

Non metals such as nitrogen, carbon, sulfur, iodine etc. doping of TiO$_2$ have shown great potential in the development in the new visible light active photocatalyst [9]. Nitrogen is one of the most commonly use for the doping because of its high stability and small ionization energy. Moreover, its comparable atomic size with oxygen can be easily introduced in the TiO_2 structure. To the best of our knowledge, Asahi et al. [10] was the very first group applied the N-TiO_2 ($TiO_{2-x}N_x$) as a visible light photocatalytic degradation of methylene blue and gaseous acetaldehyde. Since then, a lot of research works have been published on the synthesis of N-TiO_2 for the environmental applications. Several approaches has been used to introduce the nitrogen into TiO_2 structure such as wet chemical methods and physical methods. Physical methods such as atomic layer deposition [11], pulsed laser deposition [12, 13], sputtering [14, 15], ion implementation [16], and so forth have been applied for the doping of nitrogen into TiO_2. However, the most applied technique for N-TiO_2 synthesis is sol-gel method. The simplicity, low cost and are easy operation the foremost advantages of this method. By this method martial with controlled morphology and porosity can be easily synthesized.

Lee and coworkers [17] synthesized the nonporous N-TiO_2 by sol-gel method and ultrasound irradiation using urea as nitrogen source. The Photoluminescence results revealed that the intensity of N-TiO_2 was much lower than that of TiO_2. The low intensity of N-TiO_2 was due to the reduced electron hole recombination. The photocatalytic efficacy of N-TiO_2 for the degradation of reactive black 5 and rhodamine B was almost 100 % within 70 min visible light irradiation.

Liu et al. [18] grown the highly ordered N-doped TiO_2 nanotube arrays by electrochemical anode oxidation of Ti foil followed by treatment with nitrogen-plasma and subsequent annealed under Ar atmosphere. The XPS results demonstrated that all the incorporated N_2 was interstitial N-TiO_2. The UV-vis DRS analysis showed that the band gap energy of pure TiO_2 nanotubes and N-doped TiO_2 were ~ 3.24 and ~ 3.03 eV. The photocatalytic activity of TiO_2 nanotubes and N-TiO_2 nanotubes for the degradation of methylene blue under visible light irradiation was 68 and 98 % within 90 min, respectively. The experiments were performed by taking dye concentration 10 mg/L and 350 W Xe lamp.

Interstitial N doped TiO_2 was synthesized by sol-gel method using three different type of nitrogen precursors: diethanolamine, triethyl-amine and urea [19]. Results revealed that nitrogen source; dietha-nolamine provided the highest visible light absorption ability of

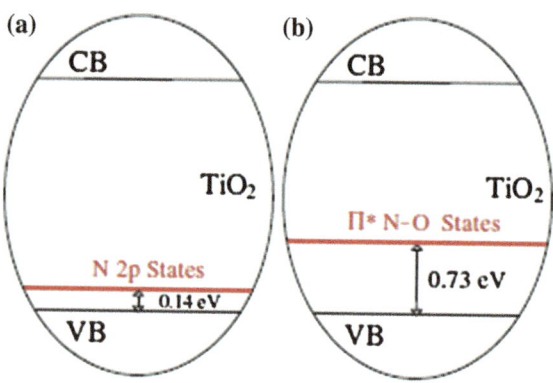

Fig. 2 States generated by substitutional (**a**) and interstitial (**b**) N-doping in TiO_2 [22]

interstitial N-TiO_2 with the lowest band gap energy (2.85 eV) and the smallest anatase crystal size (4.86 nm). N doped TiO_2 with diethanolamine showed the highest degradation of 2-chlorophenol compared to triethylamine and urea doped (N doped TiO_2).

However, researchers have different views about the mechanism of visible light photocatalytic activity of N-TiO_2. Some scientists believe that substitutional/interstitial N atoms or NO_x impurities are responsible for the higher photocatalytic activities [19, 20] while some ascribed to O vacancies are responsible for enhanced photocatalytic activity [18]. Moreover, O vacancies in N-TiO_2 could results in donor states located below the CB which enable photoreactions under visible light irradiation [18, 21, 22]. Both the localized states for N doping in TiO_2 is shown in Fig. 2.

To identify the N doping as substitutional element on the O lattice or at interstitial lattice sites, X-ray photoelectron spectroscopy (XPS) is a power full tool to discriminate the two sites. The binding energy of N 1 s at around 397 eV is the representative of substitutional N, while the peaks at >400 eV are assigned to NO (400 eV) and NO_2 (406 eV) reveals the interstitial N [19, 23]. Literature review revealed that doping method or technique is very important factor for the synthesis of N doped TiO_2 which may influence the state of nitrogen in TiO_2 and its efficiency in the visible light absorption.

The formation of oxygen vacancies in the bulk usually due to the N doping into the TiO_2 lattice, could act as recombination centers for carriers resulted in reduction the photocatalytic efficiency [24].

Carbon doping of semiconductor photocatalyst is most growing field getting significant attention of researchers. Doping of C in TiO_2 structure generate new C 2p state close to the O 2p band edge of the TiO_2 which shift absorption of light from UV to visible region [25]. Moreover, porosity and the high surface area due to C doping is another factor for enhanced photocatalytic activity. The higher surface area and porous surface facilitates the more active sites for adsorption of the pollutant ions or molecules [25, 26]. Neti et al. [27] synthesized the C doped TiO_2 using resorcinol-formaldehyde (RF) gel formation in situ during hydrolysis of Ti precursor. The band gap energy of C-TiO_2 was 2.88. The photocatalytic activity of C-TiO_2 for 2-chlorophenol in simulated solar and sunlight were found to be 56 and 36 %, respectively.

The ionic radius of the anionic dopant is most prominent factor to replace the oxygen from the TiO_2 lattice. Similar or similar size anionic dopant can be easily replace the oxygen and create the defects in TiO_2 to make it visible light active.

Similar to the N and C doping, fluorine has been successful doped in TiO_2 structure [28–30]. The size of F atom is 1.33 Å (F^-) has almost similar ionic radius and could easily replace O atoms (1.4 Å O^{2-}) [28, 29]. F doping into TiO_2 does not shift the band gap, however it improve surface acidity and causes formation of reduced Ti^{3+} ions due to the charge compensation between F^- and Ti^{4+n}, resulting in the reduction in electron hole pair recombination and higher photocatalytic degradation [25, 31]. He and coworkers [32] fluorinated anatase TiO_2 nanosheet for the oxidation of 2 chlorophenol and reduction of Cr(VI) under visible light. The results demonstrated that surface fluorination promote the formation of O vacancies and unsaturated titanium atoms which enhance the lifetime of photogenerated election-hole pairs and produced oxidizing species for 2 chlorophenol degradation and Cr(VI) reduction.

Doping of nonmetals into TiO_2 structure is depending on the ionic radius of the dopant and O. The ionic radius of sulfur (1.8 Å) is much higher than the N (1.71 Å), F (1.33 Å) and O (1.4 Å) [33]. Therefore, doping of S is comparatively more difficult. Incorporation of anionic sulfur (S^{2-}) is more difficult compared to cationic sulfur (S^{6+}). The insertion of an S 3p band below the CB shift the band gap energy and ensure the enhanced photocatalytic activity [25]. Chaudhuri and Paria [34] synthesized the S doped hollows TiO_2 nanocatalyst and investigate its application for methylene blue degradation under solar light. Hollow $S-TiO_2$ has high surface area and low band gap energy (318.11 m^2 g^{-1} and 2.5 eV) compared to solid particles of TiO_2. The $S-TiO_2$ nanocatalyst shows 98.6 % degradation of methylene blue and able to degrade 71 % of dye after five cycles. These results showed that $S-TiO_2$ is a highly efficient photocatalyst for the degradation of organic pollutant.

In recent years, non metal co-doping has been explored for the further enhance the visible light photocatalysis [35, 36]. Co-doping could modify the surface characteristics any introduce new photoactive centers, and tuning the band structure. Mani et al. C and N co-doped TiO_2 with the high surface area for the photocatalytic degradation of methyl orange under visible light irradiation. The band gap energy of $C/N-TiO_2$ was found 3.0 eV. Liu and coworkers [36] prepared the F and N co-doped TiO_2 for the degradation of rhodamine B. Upon annealing, the texture property and surface property were changed. The defect density plays the important role in the visible light absorption and the $F/N-TiO_2$ sample annealed at 300 °C showed the significant photodegradation of dye. N,S-co-doped TiO_2 photocatalyst was synthesized by the hydrolysis $TiCl_4$ using ammonia in the presence of glacial acetic acid and ammonium sulfate [37]. The photocatalytic efficiency of $N/S-TiO_2$ for phenol was much higher the commercial Degussa P25. N and S doping not only reduces its band gap but also improve the anatase crystallization and thermal stability. The high photocatalytic activity of the $N/S-TiO_2$ can be credited to the results of the synergetic effects of its high surface area, large pore volume, well-crystallized anatase, red shift in adsorption edge and strong absorbance of light with longer wavelength.

5 Metal Ions Doping (Cationic Doping)

Doping of metal ions into the TiO_2 structure (Cationic doping) has demonstrated the efficient method for enhance the visible light absorption. The electrons injected into CB from VB can be easily transported to surface of the catalyst and electrons can be easily trapped by doping with transition metals, noble metals, poor metals, earth metals [25]. Doping of metallic cations enhance the redox potential of the radicals generated during photocatalysis and reduce the e^-/h^+ recombination life.

Various transition metal ions such as Ni(II), Cr(III) Fe(III) and Zn (II) have been investigated as dopant with TiO_2 to reduce the bath chromic shift. This red shift of band edge of TiO_2 is probably due to the overlapping of conduction band Ti(3d) with d levels of the transition metals which allow absorption of light into the visible region [38]. The ionic radius of transition metals such as Fe^{3+}(0.69 Å), Cr^{3+}(0.76 Å), Ni^{2+}(0.72 Å), and Zn^{2+}(0.74 Å) are quite closer to the ionic radii of $Ti4^+$(0.75 Å). Therefore incorporation of these transition metals into TiO_2 lattice is much easy [38, 39]. However, doping of transition metals ions may also reduce the quantum efficiency because it may act as recombination sites for the photogenerated charge carriers. Moreover, transition metal doping may also make anatase TiO_2 thermally unstable [40]. The photocatalytic activity may vary according to the nature, dopant and amount of the doping agent. Photo corrosive nature of the dopant also be a limiting factor in the photocatalytic reactions.

Jaimy et al. [41] used sol gel method for the doping of Cr(III) into the TiO_2 lattice. They observed that transformation of anatase TiO_2 to rutile TiO_2 was suppressed up to 800 °C by Cr(III) ions doping. Due to doping, a red shift in the band gap energy was observed and 0.25 mol% Cr(III) doped TiO_2 calcinated at 800 °C showed the highest photocatalytic degradation of the methylene blue. The following steps were expected to be involved in the photocatalytic reaction.

$$Cr(III) + h\nu \rightarrow Cr(IV) + e^- \qquad (7)$$

$$e^- + O_2 \rightarrow O_2^- \qquad (8)$$

$$Cr(IV) + OH^- \rightarrow OH^\bullet + Cr(III) \qquad (9)$$

Chen et al. [42] compared the photocatalytic activity of Fe(III), Ni (II), Cu(II) and Co(II) doped TiO_2. Hydrothermal methd was used for the doping of transition metals. Results demonstrated that the Cu(II) and Co(II) doped TiO_2 shows the highest activity for methylene blue dye under both UV and visible light.

Manganese ion doped TiO_2 powders and thin films were prepared using sol-get method by Sharma and coworkers [43]. The dopant concentrations were varied from 2–10 mol%. The synthesized materials were used for the degradation of methylene blue and the results demonstrated that synthesized materials were more active in UV light than the visible. The possible path for the photocatalysis reaction on Mn-TiO_2 was suggested by the authors

$$TiO_2 + h\nu \rightarrow h_{vb}^+ + e_{CB}^- \qquad (10)$$

$$Ti(IV) + e_{CB}^- \rightarrow Ti(III) \qquad (11)$$

$$O^{2-} + h\nu \rightarrow O^- + e^- \qquad (12)$$

$$Mn(II) + e_{CB}^- \rightarrow Mn(I) \qquad (13)$$

$$Mn(I) + O_2 \rightarrow Mn(II) + O_2^{\bullet-} \qquad (14)$$

$$Mn(II) + h_{vb}^+ \rightarrow Mn(III) \qquad (15)$$

$$Mn(III) + OH^- \rightarrow Mn(II) + OH^\bullet \qquad (16)$$

Alternatively, the Mn(III) can also trap CB electron or Mn(I) can trap VB hole to retain half filled electronic structure of Mn(II)

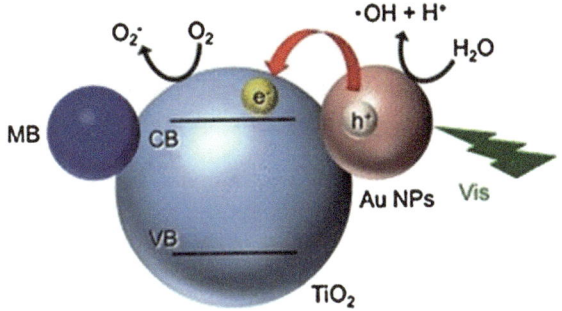

Fig. 3 Visible light photocatalytic degradation of Methylene blue onto Au/TiO$_2$ [47]

$$Mn(I) + h_{vb}^+ \rightarrow Mn(II) \qquad (17)$$

$$Mn(III) + e_{CB}^- \rightarrow Mn(II) \qquad (18)$$

$$Mn(II) + Ti(III) \rightarrow Mn(I) + Ti(IV) \quad \text{(electron trap)} \qquad (19)$$

$$Mn(II) + O^- \rightarrow Mn(III) + O^{2-} \quad \text{(hole trap)} \qquad (20)$$

The use of the noble metals such as silver (Ag), gold (Au), ruthernium (Ru), platinum (Pt), osmium (Os) etc. have been investigated as dopent to red shift the band gap energy from UV to visible light. The metals are resistance to photocorrosion and oxidation. The noble metal act as electron scavenger in charge separations and visible light absorbing sensitizer [44]. Ag doped TiO$_2$ nanomaterials were synthesized and used a visible light photocatalyst for the rhodhamine 6G by Seery [45]. The results revealed that as the amount of silver doing increases, the rate of dye degradation increases. Kuo et al. [46] prepared the Pt and Ag doped TiO$_2$ coating thin film photreactor using photoreduction process. The physic-chemical studies reveled that Ag-TiO$_2$ thin film consists Ag$_2$O, Ag and TiO$_2$ while Pt-TiO$_2$ coating contain Pt and TiO$_2$. The photocatalytic results for o-cresol degradation revealed that Pt-TiO$_2$ coating showed the higher degradation rate compared to the Ag-TiO$_2$ coating. Absorption of visible light and electron transfer from Au to the CB of TiO$_2$ is shown in Fig. 3 [47].

Jiao and coworkers [48] synthesized the Ru doped TiO_2 film and observed the better photocatalytic efficiency for the phenol degradation. The results demonstrated that the doping act as the separation of photogenerated electrons and induced the electron transfer rate at the surface.

Application of rare earth metals scandium (Sc), yttrium (Y), cerium (Ce), gadolinium (Gd) etc. have been investigated as dopant due to its incompletely filled 4f and empty 5d orbitals [26]. Stengl and coworkers [49] prepared investigate the photocatalytic activity of La, Ce, Pr, Nd, Sm, Eu, Dy, Gd doped TiO_2 for the decomposition of orange second dye. The photocatalytic activity with Nd(II) and Ce (IV) doping were much higher. The transitions of 4f electrons of rare earth led to the enforcement of the optical adsorption of catalysts and enhanced the life photo-generated electron–hole pairs. Moreover the photocatalytic activity of the catalyst also increase with the increase in the doping content and the 0.5–1.0 % rare earth ion doping shows the highest activity while in case on Nd(III) doing maximum degradation was observed at 10 wt% doing. A series of Er(III), Yb(III) and Er(III)/Yb(III) co-doped TiO_2 photocatalysts were synthesized via sol-gel method [50]. The XPS results demonstrate that terbium and ytterbium were present in the oxide form. Yb(III)-TiO_2 containing 1 mol% Yb(III) showed the highest photocatalytic activity for phenol degradation. The results demonstrated that the visible light sensitization was probably due to the increase in surface area, reduction in crystallite size, increase in adsorption sites and deterrence of electron–hole recombination. XRD analysis indicates incorporation of rare earth metal ions into TiO_2 lattice creating oxygen vacancies and surface defects.

Poor metals such as tin (Sn), aluminum (Al), lead (Pb), gallium (Ga), indium (In), thalium (TI), and bismuth (Bi) are the p block elements having lower melting and boiling point than transition metals. Tsai et al. [51] synthesized Al doped TiO_2 by direct combination of vaporized Ti, Al, and O_2 using a 6 kW thermal plasma system. Both rutile and anatase phases were present in the Al-TiO_2. Ti (III) and Ti(IV) forms were coexisted in Al doped TiO_2 and the amount of Ti(III) increased with the increase in the Al concentrations due to Al/Ti substitution which caused the occurrence of O vacancies. Shaoyou et al. [52] reported that small amount of Al doping in TiO_2 photocatalytic degradation of congo red dye and which was much higher than that of pure TiO_2. The photocatalytic performances of Bi-TiO_2 were studied by degradation of 2,4-dichlorophenol (2,4-DCP) in transparent aqueous solutions under visible light illumination [53]. Results revealed that mesoporous polyethylene glycol-modified Bi-TiO_2 exhibited much higher photocatalytic activities than Bi-TiO_2 without modification, and noticeably the optimized doping level was increased from 2 to 4 mol%. The higher activity of PEG-modified Bi-TiO_2 could be attributed to well mix crystal phase, high surface area and porosity, and its strong absorption in the visible region. The introduction of Bi induces the Bi(IV)/Bi(III) species which act as electron trap and thus ensure the separation of the electron–hole pairs [54]. Sui and coworker [55] reported the effectiveness of the Sn doped TiO_2. Doing of Sn(IV) prevent the phase transformation of TiO_2 from anatase to rutile, decreases the diameter of TiO_2 photocatalyst, and enhances significantly the photocatalytic activity. The effectiveness of Sn into TiO_2 lattices was due to similar ionic radii of Sn(IV) (0.71 Å) and Ti(IV) (0.68 Å).

6 Coupled, Composite and Hybrid TiO_2 Photocatalysts

Modification of photocatalyst by coupling between two or more semiconductors with different energy levels is an additional approach to make the visible light active photocatalyst [33, 56, 57]. When a semiconductor having low band gap energy and negative conduction band level is coupled with the large band gap of TiO_2, the electrons can be transferred from low band gap of the semiconductor to the TiO_2. The interfacial charge transfer processes are influenced by the sensitizer semiconductor and promote the separation of photogenerated electrons and holes [25]. Bessekhouad et al. reported that the transfer of photogenerated electrons and holes between the low band gap sensitizer and TiO_2 mainly depends on the difference between the CB and VB potentials of the applied semiconductors [57]. Upon the visible irradiation, only sensitizer semiconductor absorbs the visible light photons and get excited. These excited electrons flow into the CB of the closest TiO_2 [58]. Negatively charged CdS quantum dots were evenly deposited on the TiO_2 nanotubes arrays [59]. The CdS-TiO_2 nanotube array hybrid nanostructures show efficient visible-light photoactivity towards the photo oxidation of methyl orange and the photocatalytic reduction of nitrophenol derivatives. CdS upon visible light irradiation, photo excited electron from CB of CdS injected to the CB of the TiO_2 nanotubes. These injected electrons radically transferred to the bottom of Ti foil and enhance the electron hole charge separation and produced the highly active oxygen species responsible for the degradation of dye. While the photogenerated holes selectively reduce the nitrophenol derivatives. However, photocorrosion is very common in CdS due to oxidation by photogenerated holes. In CdS-TiO_2, surrounding TiO_2 matrix protect the CdS against photocorrosion [60]. In CdS-TiO_2 nanotube array hybrid nanostructures, photogenerated holes can be transfer to adsorbed water on the TiO_2 surface, which prevent holes to react with CdS directly [59]. Liu et al. [61] fabricated CdS/TiO_2 nanowires on the surface zeolite by sol-gel and hydrothermal synthesis method. Figure 4, shows the generation of oxidizing and reduction species on the surface of composite. In this composite zeolite act as an adsorbent and CdS/TiO_2 degrade the methyl blue though the photocatalysis process. V_2O_5 semiconductor is another low band gap energy (2.3 eV) material used for the coupling with the TiO_2. Wang et al. [62] prepared the V_2O_5/TiO_2 nanostructure by electrospinning and observed that 1:1 molar ration of V/Ti showed the highest visible light photocatalytic activity for rhodamine B dye even after 10 regeneration cycles.

Libera et al. [63] reported the photocatalytic activity of the Fe_xO_y centers created on the surface of TiO_2 using Atomic Layer deposition technique. Authors reported that the TiO_2 and a physically mixed Fe_2O_3/TiO_2 did not show the any significant decomposition of methylene blue compared to Fe_xO_y/TiO_2 under visible light. The

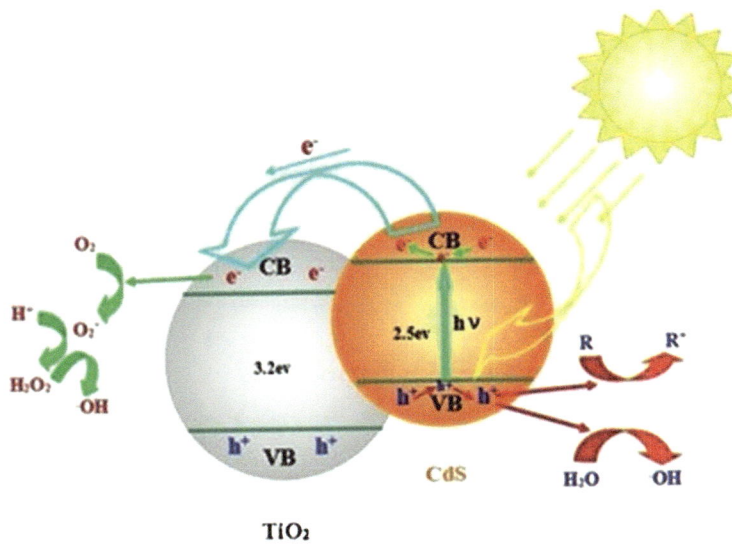

Fig. 4 Photocatalytic degradation mechanism for CdS/TiO₂/zeolite composite materials [61]

higher efficacy of Fe_xO_y/TiO_2 was due to the efficient charge separation and unique redox properties of photogenerated changes.

WO_3 is well known semiconductor and TiO_2/WO_3 is a coupled semiconductor-semiconductors system which has received great attention of the researchers in the field of photocatalysis [6, 25, 64]. Under the visible light irradiation, WO_3 can be excited and electron from the WO_3 can be transferred to the surface adsorbed O_2 and holes can be transferred from WO_3 to TiO_2. Lu and coworkers [64] synthesized the WO_3 decorated on the internal external sidewall of TiO_2 nanotubes ($WO_3@TiO_2@WO_3$). The UV and visible light photocatalytic activity of TiO_2 nanofibers, TiO_2 nanotubes, and TiO_2 nanofibers decorated with WO_3 nanoparticles, were prepared in order to compare with the $WO_3@TiO_2@WO_3$. The heterostructure $WO_3@TiO_2@WO_3$ showed a wide range of light absorption and demonstrated the highest photocatalytic activity for rhodamine B dye. The higher photocatalytic degradation rates of organic pollutants could be mainly due to the increase in the photogenerated electron transfer. Higher rates of pollutant degradation were recorded owing to

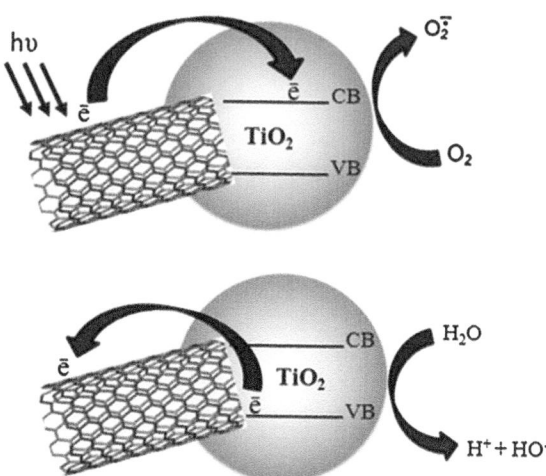

Fig. 5 Processes in semiconductor photocatalysis. Photoabsorption and electron–hole pair generation from change of separation and migration of the electron to the surface [68]

the increase of transfer from TiO_2 to WO_3 via the formation of an intermediate W(V) species, together with the increased surface acidity in the TiO_2/WO_3 couple [25, 65].

In recent few years, nanocarbon based photocatalysts has been widely explored in photocatalytic applications. Graphene and carbon nanotubes belong to this class. Both are considered as good supports for semiconductors with photocatalytic properties because of their high surface area, porous structure, high mechanical strength and chemical stability. Combination of graphene (GN) and CNTs with TiO_2 as composite photocatalyst for the degradation of organic pollutants has been investigated. GN and CNTs allow a efficient and faster electron and heat transfer [66]. These nanocarbon can efficiently adsorb the organic pollutant in aqueous solution and also promote the photocatalytic activity of the TiO_2 by trapping the photogenerated electrons and enhance the charge separation [66, 67]. The movement of the electron in CNT/TiO_2 composite has been shown in Fig. 5 [68].

Kuvarega et al. [67] synthesized N and Pd co-doped MWCNTs/TiO_2 nancomposite for the degradation of eosin yellow

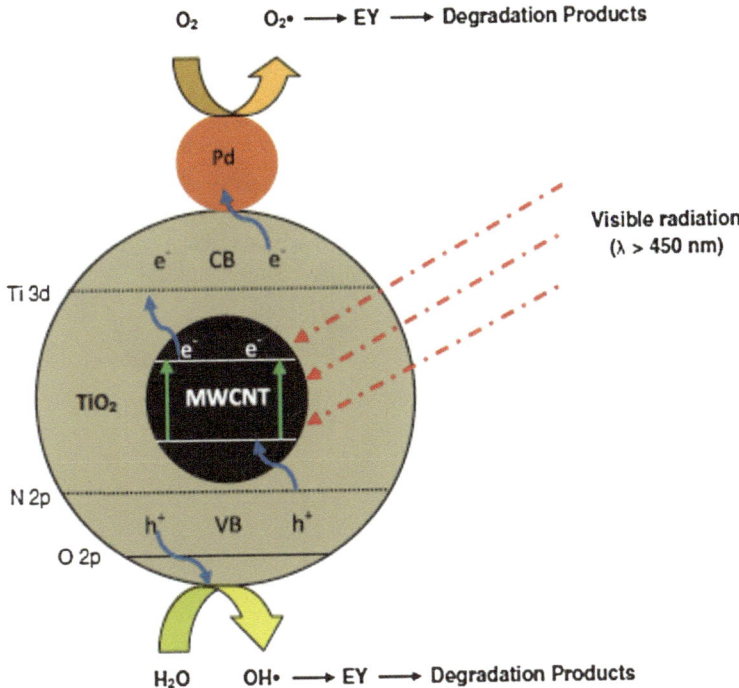

Fig. 6 The excitonic process and degradation of dye eosin yellow onto N and Pd co-doped MWCNTs/TiO₂ nanocomposite [67]

under simulated and solar light irradiation. Figure 6 shows the excitonic process and degradation of dye onto N and Pd co-doped MWCNTs/TiO₂ nanocomposite. The authors reported that 0.5 wt% content in the composite shows the best photocatalytic results. The higher MWCNTs loading in composite reduces the degradation of dye. In another study, Wang et al. reported that 5 wt% loading of MWCNTs in MWCNTs/TiO₂ composite showed the maximum degradation of 2,6 dinitro-p-cresol [69].

Similar to CNTs, nanocomposite of GN (or GR) with TiO₂ shown the enhanced photocatalytic property. GN also work as an acceptor material due to its π-conjugation structure, thus in TiO₂/GN system, the excited electrons of TiO₂ could transfer from the conduction band to GN. Moreover, two dimantional planer structure and π-π interaction between composite and organic pollutant are the main reasons for

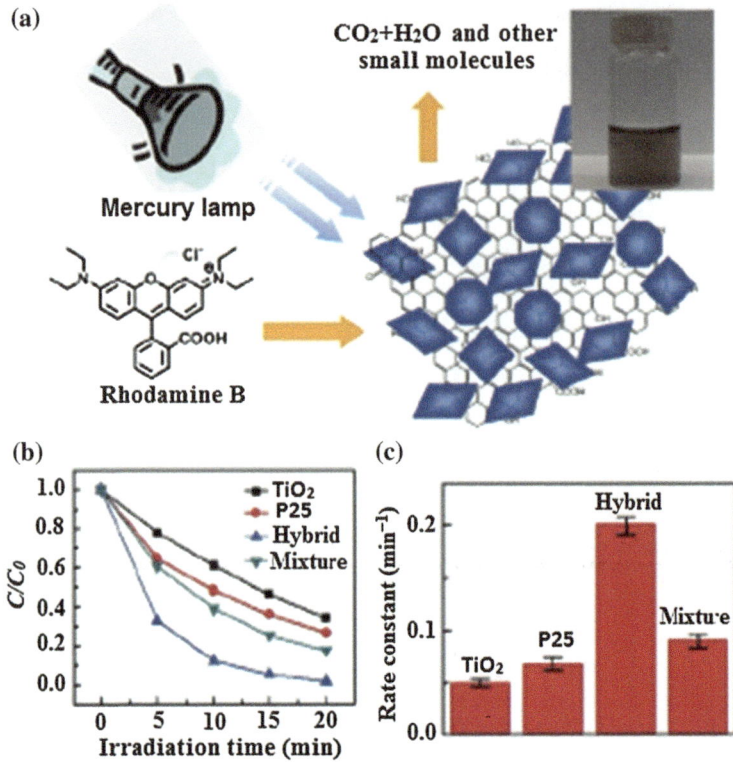

Fig. 7 a Schematic illustration of the photodegradation of rhodamine B molecules by the graphene/TiO_2 nanocrystals hybrid under irradiation by a mercury lamp. The inset shows the solution of the graphene/TiO_2 nanocrystals hybrid. **b** Photocatalytic degradation of rhodamine B monitored as normalized concentration change versus irradiation time in the presence of free TiO_2, P25, graphene/TiO_2 nanocrystals hybrid and a graphene/TiO_2 mixture. **c** Average reaction rate constant (min^{-1}) for the photodegradation of rhodamine B with free TiO_2, P25, graphene/TiO_2 nanocrystals hybrid, and the graphene/TiO_2 mixture. The error bars are based on measurements on at least four different samples [71]

the enhanced photocatalytic activity of the composite catalyst [66, 70]. Graphene oxide/TiO_2 nanocrystal hybrid as a superior photo-catalyst has been synthesized by direct gropwn of nanocrystal on GO sheet [71]. Synthesized GO/TiO_2 nanocrystal hybrid showed three fold photocatalytic degradation rhodamine B compared to P25 as shown in Fig. 7. Muthirulann [72] showed the electron transfer mechanism from CB of TiO$_2$ to GR as follows:

$$TiO_2/GR + h\nu \rightarrow TiO_2(h^+)-GR(e^-) \qquad (21)$$

$$GR(e^-) + O_2 \rightarrow GR + O_2^- \qquad (22)$$

$$TiO_2(h^+) + H_2O/OH \rightarrow TiO_2 + {}^\bullet OH \qquad (23)$$

$${}^\bullet OH + acid\ orange\ 7 \rightarrow degradation\ products \qquad (24)$$

Ternary PbS-graphene/TiO$_2$ composites were synthesized by the sol-gel method by Ullah et al. [73]. The authors observed that the activity of the composite in visible region was due to the coupling of PbS with TiO$_2$ while graphene act a photogenerated charge separator and transporter. Ullah et al. [73] proposed the following mechanism for the degradation of methylene blue under visible light irradiation.

$$PbS - TiO_2 + h\nu \rightarrow PbS(h^+, e^-) - TiO_2 \qquad (25)$$

$$PbS(h^+, e^-) - TiO_2 \rightarrow PbS(h^+) + TiO_2(e^-) \qquad (26)$$

$$e^- + O_2 \rightarrow O_2^{\bullet-} \qquad (27)$$

$$h^+ + H_2O \rightarrow {}^\bullet OH + H^+ \qquad (28)$$

$$O_2^{\bullet-}\ or\ {}^\bullet OH + methylene\ blue \rightarrow CO_2, H_2O \qquad (29)$$

In recent years, graphitic carbon nitride (g-C$_3$N$_4$) has caught the much attention due to its promising visible light photocatalytic application [74]. The highest occupied molecular orbital of the g-C$_3$N$_4$ (−1.12 eV) is more negative than TiO$_2$. Therefore, g-C$_3$N$_4$ can be coupled with the TiO$_2$ to form the new vsisble light photocatalyst due to closer band gap edge potential [75]. Zhu et al. [76] synthesized the g-C$_3$N$_4$-TiO$_2$ composite to improve the visible light absorption and visible light photocatalytic activity of TiO$_2$. The synthesize composite shows the higher photocatalytic activity then the pure g-C$_3$N$_4$ under visible light irradiation. The enhanced visible light activity was mainly due the formation of heterojunction which enhanced the charge

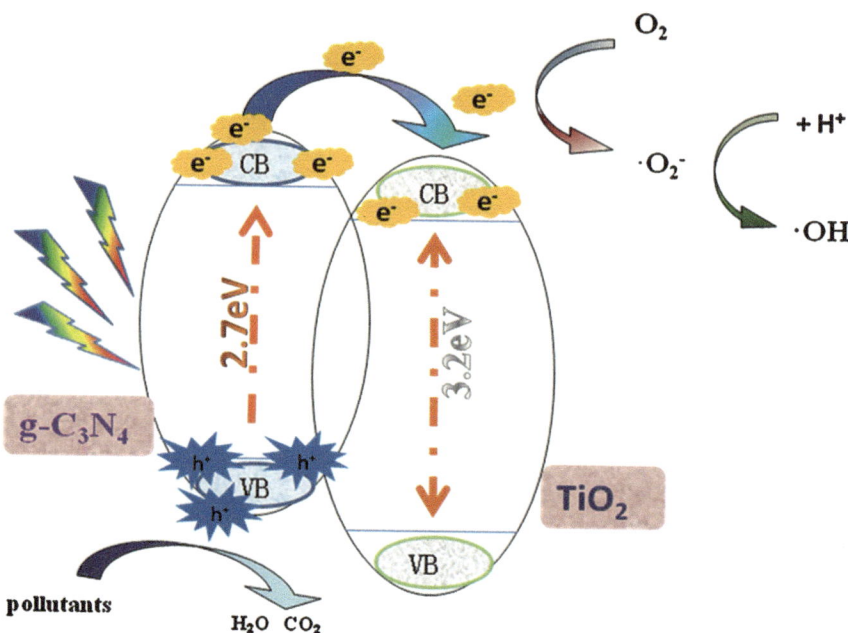

Fig. 8 Proposed mechanism for the photoexcited electron–hole separation and transport processes at the g-C$_3$N$_4$/TiO$_2$ interface under visible light irradiation [76]

separation. The electron spin resonance results demonstrate that O$_2$$^{\bullet-}$ radicals was the mainly produced by g-C$_3$N$_4$-TiO$_2$ and g-C$_3$N$_4$. A schematic illustration for g-C$_3$N$_4$-TiO$_2$ photocatalytic mechanism is shown in Fig. 8.

Utilization of conducting polymer as a photosensitizer is also a good approach to improve the visible light photocatalytic activity of TiO$_2$. These conducting polymers are able to absorb visisble light and inject the elections to the CB of the TiO$_2$ [77, 78]. These conduction band electrons subsequently transport to the surface of photocatalyst to react with absorbed O$_2$ and H$_2$O to generate $^{\bullet}$O$_{2-}$ and $^{\bullet}$OH. Polythiophenes [79, 80], polypyrroles [81, 82], polyanilines [83–85], [83] and so on [78, 86] have been widely conjugated with TiO$_2$. The conducting polymer/TiO$_2$ photocatalysts not only enhance the visible light photocatalytic activity but also facilitating the separation of photogenerated carriers.

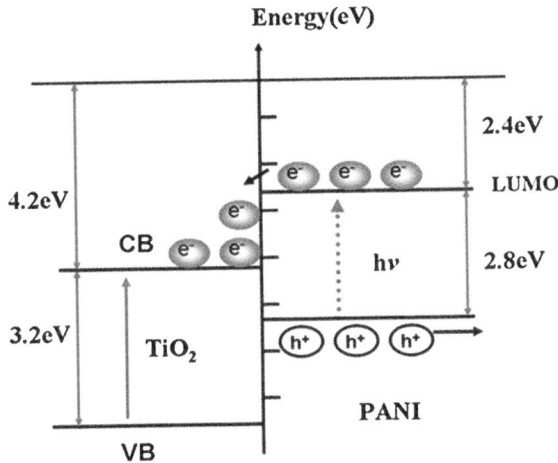

Fig. 9 Photocatalytic mechanism for PANI-TiO$_2$

Wai et al. [83] synthesized the Polyaniline (PANI)-TiO$_2$ nano-composite though a facile hydrothermal route for the photocatalytic degradation of acetone. The synergetic effect between PANI and TiO$_2$ facilaitas the better degradation and prevent charge recombination. Figure 9 shows photocatalytic mechanism for PANI-TiO$_2$ shows the role of PANI and photocatalytic degradation mechanism of PANI-TiO$_2$. Upon visisble light illumination, PANI-TiO$_2$, the PANI absorb photons to induce π-π* trasition and simultaneously electrons injected to the CB of TiO$_2$. The photogenerated electron and holes transferred to the surface and produce hydroxyl and superoxide radicals after reacting with H$_2$O and O$_2$. These radicals degrade the organic molecules.

Xu et al. [79] studied the adsorption and photocatalytic decomposition of methyl orange by polythiophene/titanium dioxide (PTh/TiO$_2$) composite. The results demonstrated the dye can be mineralized to the CO$_2$ and H$_2$O with some intermediate products under visisble light photocatalysis. Moreover the photocatalysis of methyl orange increases with the increase illumination time and low concentration of dye favored the degradation rate. Fang et al. developed a reduced GO/TiO$_2$ composite on the surface of amino-poly (styrene-co-glycidyl methacrylate (TiO$_2$/rGO/polymer).

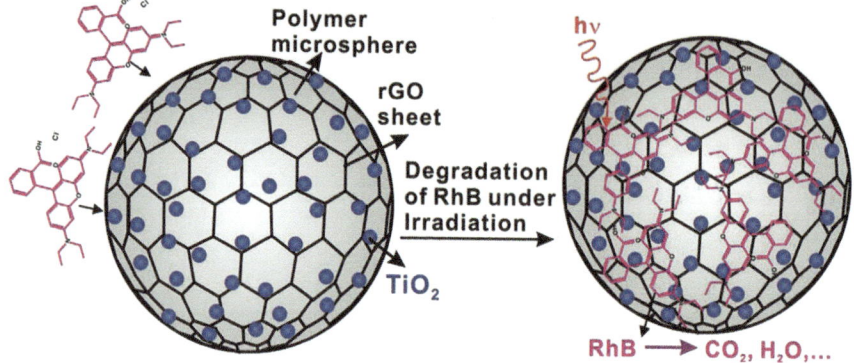

Fig. 10 Proposed mechanism for the photocatalytic degradation of rhodamine B by (TiO_2/rGO/polymer) composite under visible light [86]

The authors explain the photocatalytic mechanism on the basis of dye sensitization (Fig. 10). Under visible light irradiation, dye molecules can absorb the photons and get excited. The excited electron transferred through a chain as follows:

$$RhB + h\nu \rightarrow RhB(e^- + h^+) \tag{30}$$

$$RhB(e^-) + Graphene \rightarrow RhB(h^+) + Graphene(e^-) \tag{31}$$

$$Graphene(e^-) + TiO_2 \rightarrow Graphene + TiO_2(e^-) \tag{32}$$

$$TiO_2(e^-) + O_2 \rightarrow O_2^{\bullet-} + TiO_2 \tag{33}$$

$$O_2^{\bullet-} + RhB(h^+) \rightarrow CO_2 + H_2O + \text{other mineralization product} \tag{34}$$

7 Conclusions

This chapter highlights the recent literature that provides the insights on the modification of the TiO_2 catalyst. A large number of research works have been conducted on the development of a new TiO_2

photocatalyst ablity to absorb visible light as a main part of solar spectrum. The major boundary of pure TiO_2 is its activity only in UV light and fast recombination of photogenerated electrons and holes recombination rate. Current retrench reports that hurdle can be overcome by introducing foreign species into the titanium structure or onto TiO_2 surface. This chapter deals with the modification of TiO_2 and its photon absorption, charge transfer and trapping ability under visible light irradiation. The applications of modified TiO_2 for the degradation of organic pollutants from aqueous solution were also discussed. Up to now, large numbers of research work have been published showing successful applications of TiO_2 photocatalyst. This most of the published works were conducted at the laboratory scale. Future research should be focused on the use of modified TiO_2 for large scale applications using solar photocatalytic systems.

References

1. M. Luisa Marin, Lucas Santos-Juanes, Antonio Arques, Ana M. Amat, and Miguel A. Miranda. Organic Photocatalysts for the Oxidation of Pollutants and Model Compounds. Chem. Rev. 2012, 112, 1710–1750.
2. Braslavsky, S. E. Glossary of terms used in photochemistry, 3rd edition (IUPAC Recommendations 2006). Pure Appl. Chem. 2007, 79, 29–465.
3. B. Ohtani Principle of Photocatalysisand Design of Active Photocatalysts Chapter 5 in New and Future Developments in Catalysis. 2013, Pages 121–144.
4. Swagata Banerjee, Suresh C. Pillai, Polycarpos Falaras, Kevin E. O'Shea, John A. Byrne, and Dionysios D. Dionysiou. New Insights into the Mechanism of Visible Light Photocatalysis. J. Phys. Chem. Lett. 2014, 5, 2543–2554.
5. Fujishima, A.; Honda, K. Electrochemical Photolysis of Water at a Semiconductor Electrode. Nature 1972, 238, 37–38.
6. Shipra Mital Gupta, Manoj Tripathi. A review of TiO2 nanoparticles. Chinese Sci Bull June (2011) 56, 1639–1657.
7. Cassaignon, C. Colbeau-Justin and O. Durupthy . Chapter 6 Titanium Dioxide in Photocatalysis. In, R. Brayner et al. (eds.), Nanomaterials: A Danger or a Promise?, DOI: 10.1007/978-1-4471-4213-3_6, _ Springer-Verlag London 2013S.
8. J. Rashid, M. A. Barakat, S. L. Pettit & J. N. Kuhn (2014) InVO 4/TiO 2 composite for visiblelight photocatalytic degradation of 2-chlorophenol in wastewater, Environmental Technology, 35:17, 2153–2159.

9. M. Pelaez, N.T. Nolan, S. C. Pillai, M. K. Seery, P. Falaras, A. G. Kontos, P. S. M. Dunlop, J.W.J. Hamilton, J. A. Byrne, K. O. Shea, M. H. Entezari, D. D. Dionysiou . A review on the visible light active titanium dioxide photocatalysts for environmental applications. Applied Catalysis B: Environmental 125 (2012) 331–349.

10. R. Asahi, T. Morikawa, T. Ohwaki, K. Aoki, Y. Taga, Visible-light photocatalysis in nitrogen-doped titanium oxides. Science 293 (2001) 269–271.

11. V. Pore, M. Heikkila, M. Ritala, M. Leskela, S. Arev, Atomic layer deposition of $TiO2_{-x}N_x$ thin films for photocatalytic applications Journal of Photochemistry and Photobiology A: Chemistry 177 (2006) 68–75.

12. G. Socol, Yu. Gnatyuk, N. Stefan, N. Smirnova, V. Djokić, C. Sutan, V. Malinovschi, A. Stanculescu, O. Korduban, I.N. Mihailescu. Photocatalytic activity of pulsed laser deposited TiO2 thin films in N_2, O_2 and CH_4. Thin Solid Films, 2010; 518, 4648–4653.

13. L. Zhao, Q. Jiang, J. Lian, Visible-light photocatalytic activity of nitrogen-doped TiO2 thin film prepared by pulsed laser deposition Applied Surface Science 254 (2008) 4620–4625.

14. S-H. Lee, E. Yamasue, H. Okumura, K.N. Ishihara, Effect of oxygen and nitrogen concentration of nitrogen doped TiO_x film as photocatalyst prepared by reactive sputtering. Applied Catalysis A: General 371 (2009) 179–190.

15. G. Abadias, F. Paumier, D. Eyidi, P. Guerin, T. Girardeau, Structure and properties of nitrogen-doped titanium dioxide thin films produced by reactive magnetron sputtering Surface and Interface Analysis 42 (2010) 970–973.

16. Li Jinlong, M. Xinxin, S. Mingren, X. Li, S. Zhenlun, Fabrication of nitrogen-doped mesoporous TiO2 layer with higher visible photocatalytic activity by plasma-based ion implantation. Thin Solid Films 519 (2010) 101–105.

17. H.U. Lee, S. C. Lee, S. Choi, B. Son, S. M. Lee, H. J. Kim, J. Lee. Efficient visible-light induced photocatalysis on nanoporous nitrogen-doped titanium dioxide catalysts. Chemical Engineering Journal 228 (2013) 756–764.

18. X. Liu, Z. Liu, J. Zheng, X. Yan, D. Li, S. Chen, W. Chu. Characteristics of N-doped TiO2 nanotube arrays by N2-plasma for visible light-driven photocatalysis. Journal of Alloys and Compounds 509 (2011) 9970–9976.

19. J. Ananpattarachai, P. Kajitvichyanukul, S. Seraphin, Visible light absorption ability and photocatalytic oxidation activity of various interstitial N-doped TiO2 prepared from different nitrogen dopants. J Hazard Mater. 2009 Aug 30;168(1):253–61.

20. J.J. Xu, Y.H. Ao, M.D. Chen, D.G. Fu, Photoelectrochemical property and photocatalytic activity of N-doped TiO2 nanotube arraysAppl. Surf. Sci. 256 (2010) 4397–4401.

21. T. Ihara, M. Miyoshi, Y. Iriyama, O. Matsumoto, S. Sugihara. Visible-light-active titanium oxide photocatalyst realized by an oxygen-deficient structure and by nitrogen doping Appl. Catal. B: Environ. 42 (2003) 403–409.

22. Ying Yang, Hui Zhong, Congxue Tian. Photocatalytic mechanisms of modified titania under visible light. Res Chem Intermed (2011) 37:91–102.

23. N.T. Nolan, D.W. Synnott, M.K. Seery, S.J. Hinder, A. Van Wassenhoven, S.C. Pillai, Effect of N-doping on the photocatalytic activity of sol–gel TiO2 Journal of Hazardous Materials 211–212 (2012) 88–94.

24. Zhang, J.; Wu, Y.; Xing, M.; Khan Leghari, S. A.; Sajjad, S. Development of modified N-doped TiO2 photocatalyst with metals, nonmetals, and metal oxides. Energy Environ. Sci. 2010, 3, 715–726.

25. R. Daghrir, P. Drogui, D. Robert. Modified TiO2 For Environmental Photocatalytic Applications: A Review. | Ind. Eng. Chem. Res. 2013, 52, 3581–3599.

26. Teh, C. M.; Mohamed, A. R. Role of titanium dioxide and iondoped titanium dioxide on phtotocatalytic degradation of organic pollutants (phenol compounds and dyes) in aqueous solutions: A review. J. Alloys Compd. 2011, 509, 1648 –1660.

27. N.R. Neti, R. Misra, P. K. Bera, R. Dhodapkar, S. Bakardjieva. Z. Bastl (2010) Synthesis of C-Doped TiO2 Nanoparticles by Novel Sol-Gel Polycondensation of Resorcinol with Formaldehyde for Visible-Light Photocatalysis, Synthesis and Reactivity in Inorganic, Metal-Organic, and Nano-Metal Chemistry, 40:5, 328–332.

28. Todorova, N.; Giannakopoulou, T.; Vaimakis, T.; Trapalis, C. Structure tailoring of fluorine-doped TiO2 nanostructured powders. Mater. Sci. Eng., B 2008, 152, 50–54.

29. Umebayashi, T.; Yamaki, T.; Tanaka, S.; Asai, K. Visible light induced degradation of methylene blue on S-doped TiO2. Chem. Lett. 2003, 32, 330–331.

30. Yu, J. C.; Yu, J.; Ho, W.; Jiang, Z.; Zhang, L. Effects of F doping on the photocatalytic activity and microstructures of nanocrystalline TiO2 powders. Chem. Mater. 2002, 14, 3808–3816.

31. A.M. Czoska, S. Livraghi, M. Chiesa, E. Giamello, S. Agnoli, G. Granozzi, E. Finazzi, C. Di Valentin, G. Pacchioni, Journal of Physical Chemistry C 112 (2008) 8951–8956.

32. Z. He, L. Jiang, D. Wang, J. Qiu, J. Chen, S. Song. Simultaneous Oxidation of p-Chlorophenol and Reduction of Cr(VI) on Fluorinated Anatase TiO2 Nanosheets with Dominant {001} Facets under Visible Irradiation. Ind. Eng. Chem. Res., 2015, 54, 808–818.

33. Zhang, H.; Chen, G.; Behnemann, D. W. Photo-electrocatalytic materials for environmental applications. J. Mater. Chem. 2009, 19,5089–5121. Rehman, S.; Ullah, R.; Butt, A. M.; Gohar, N. D. Strategies of making TiO2 and ZnO visible light active. J. Hazard. Mater. 2009, 170, 560–569.

34. R. G. Chaudhuri S. Paria. Visible light induced photocatalytic activity of sulfur doped hollow TiO2nanoparticles, synthesized via a novel route. Dalton Trans., 2014,43, 5526–5534.

35. A. Daya Mani, S. Muthusamy, S. Anandan & Ch. Subrahmanyam (2015) C and N doped nano-sized TiO2 for visible light photocatalytic degradation of aqueous pollutants, Journal of Experimental Nanoscience, 10:2, 115–125.
36. S. Liu, J. Yu, W. Wang, Effects of annealing on the microstructures and photoactivity of fluorinated N-doped TiO2. Physical Chemistry Chemical Physics 12 (2010)12308–12315.
37. Xu, J.-H.; Li, J.; Dai, W.-L.; Cao, Y.; Li, H.; Fan, K. Simple Fabrication of Twist-Like Helix N,S-Codoped Titania Photocatalyst with Visible-Light Response. Appl. Catal., B 2008, 79, 72−80.
38. M.A. Rauf, M.A. Meetani, S. Hisaindee. An overview on the photocatalytic degradation of azo dyes in the presence of TiO2 doped with selective transition metals. Desalination 276 (2011) 13–27.
39. L.G. Devi, N. Kottam, B.N. Murthy, S.G. Kumar, Enhanced photocatalytic activity of transition metal Mn^{2+}, Ni^{2+} and Zn^{2+} doped polycrystalline titania for the degradation of Aniline Blue under UV/solar light, Journal of Molecular Catalysis A: Chemical 328 (2010) 44–52.
40. W. Choi, A. Termin, M.R. Hoffmann, The Role of Metal Ion Dopants in Quantum-Sized TiO2: Correlation between Photoreactivity and Charge Carrier Recombination Dynamics Journal of Physical Chemistry B 98 (1994) 13669–13679.
41. K. B. Jaimy, S. Ghosh, S. Sankar, K.G.K. Warrier. An aqueous sol–gel synthesis of chromium(III) doped mesoporous titanium dioxide for visible light photocatalysis. Materials Research Bulletin 46 (2011) 914–921.
42. Y.Y. Chen, Y.B. Xie, J. Yang H.B. Cao, H. Liu, Y.Zhang. Double layered, one-pot hydrothermal synthesis of M-TiO2 (M = Fe^{3+}, Ni^{2+}, Cu^{2+} and Co^{2+}) and their application in photocatalysis. Sci China Chem (2013) 56, 1783–1789.
43. S.D. Sharma, K.K. Saini, C. Kant, C.P. Sharma, S.C. Jain, Photodegradation of dye pollutant under UV light by nano-catalyst doped titania thin films, Applied Catalysis B: Environmental 84 (2008) 233–240.
44. Yoon, J. W.; Sasaki, T.; Koshizaki, N. Dispersion of nanosized noble metals in TiO2 matrix and their photoelectrode properties. Thin Solid Films 2005, 483, 276−282.
45. M. K. Seery, R. George, P. Floris, S. C. Pillai. Silver doped titanium dioxide nanomaterials for enhanced visible light photocatalysis. Journal of Photochemistry and Photobiology A: Chemistry 189 (2007) 258–263.
46. Yu-Lin Kuo, Te-Li Su, Kai-Jen Chuang, Hua-Wei Chen & Fu-Chen Kung (2011) Preparation of platinum- and silver-incorporated TiO2 coatings in thin-film photoreactor for the photocatalytic decomposition of o-cresol, Environmental Technology, 32:15, 1799–1806.
47. Teruhisa Okuno, Go Kawamura, Hiroyuki Muto, Atsunori Matsuda Three modes of high-efficient photocatalysis using composites of TiO2 - nanocrystallite-containing mesoporous SiO_2 and Au nanoparticles. Journal of Sol-Gel Science and Technology 2015, 74, 748–755.

48. B. Jiao, Z. Xu, X. Sun, W.Cai, E. Wang, B. Jiao, H.Ji, Preparation and characterization of nanoparticle Ru:TiO2 films and their photocatalytic activity. Rare Metals 30, 2011, 254–258.

49. V. Stengl, S. Bakardjieva, N. Murafa. Preparation and photocatalytic activity of rare earth doped TiO2 nanoparticles. Materials Chemistry and Physics 114 (2009) 217–226.

50. J. Reszczynska, T. Grzyb, J.W. Sobczak, W. Lisowski, M. Gazda, B. Ohtani, A. Zaleska. Visible light activity of rare earth metal doped (Er^{3+}, Yb^{3+} or Er^{3+}/Yb^{3+}) titania photocatalysts. Applied Catalysis B: Environmental 163 (2015) 40–49.

51. Cheng-Yen Tsai, Tien-Ho Kuo, Hsing-Cheng His. Fabrication of Al-Doped TiO2 Visible-Light Photocatalyst for Low-Concentration Mercury Removal. International Journal of Photoenergy, 2012 ID 874509, doi:10.1155/2012/874509.

52. L. Shaoyou, G. Liu, Q. Feng. Al-doped TiO2 mesoporous materials: synthesis and photodegradation properties. Journal of Porous Materials 2009; 17,197–206.

53. Zhu, Ximiao; Liu, Zhang; Fang, Jianzhang; Wu, Shuxing; Xu, Wei Cheng. Synthesis and characterization of mesoporous Bi/TiO2 nanoparticles with high photocatalytic activity under visible light. Journal of Materials Research; 2013, 28, 1334–1342.

54. Li, H.; Wang, D.; Wang, P.; Fan, H.; Xie, T. Synthesis and studies of the visible light photocatalytic properties of near monodisperse Bi-doped TiO2 nanospheres. Chem. Eur. J. 2009, 15, 12521−12527.

55. Sui, R.; Young, J. L.; Berlinguette, C. P. Sol-gel synthesis of linear Sn-doped TiO2 nanostructures. J. Mater. Chem. 2010, 20, 498−503.

56. Ilieva, M.; Nakova, A.; Tsakova, V. TiO2/WO3 hybrid structures produced through a sacrificial polymer layer technique for pollutant photo- and photo-electrooxidation under ultraviolet and visible light illumination. J. Appl. Electrochem. 2012, 42, 121−129.

57. Bessekhouad, Y.; Robert, D.; Weber, J. V. Bi2S3/TiO2 and CdS/TiO2 heterojunctions as an available configuration for photocatalytic degradation of organic pollutant. J. Photochem. Photobiol. A 2004, 163, 569−580.

58. Robert, D. Photosensitization of TiO2 by MxOy and MxSy nanoparticles for heterogeneous photocatalysis applications. Catal. Today 2007, 122, 20−26.

59. Fang-Xing Xiao, Jianwei Miao, Hsin-Yi Wang and Bin LiuSelf-assembly of hierarchically ordered CdS quantum dots–TiO2 nanotube array heterostructures as efficient visible light photocatalysts for photoredox applications. J. Mater. Chem. A, 2013, 1, 12229–12238.

60. Bessekhouad, Y.; Chaoui, N.; Trzpit, M.; Ghazzal, N.; Robert, D.; Weber, J. V. UV-vis versus visible degradation of acid orange II in a coupled CdS/TiO2 semiconductors suspension. J. Photochem. Photobiol. A 2006, 183, 218−224.

61. Zhichao Liu, Zhifeng Liu, Ting Cui, Jing Zhang, Yufeng Zhao, Jianhua Han, Keying Guo, and Caputo Domenico. Preparation and Photocatalysis of Schlumbergera bridgesii-Like CdS Modified One-Dimensional TiO2 Nanowires on Zeolite. Journal of Materials Engineering and Performance (2015) 24:700–708.

62. Yuan Wang, Jiawang Zhang, Lixin Liu, Chengquan Zhu, Xueqin Liu, Qing Su. Visible light photocatalysis of V2O5/TiO2 nanoheterostructures prepared via electrospinning. Materials Letters 75 (2012) 95–98.

63. Joseph A. Libera, Jeffrey W. Elam, Norman F. Sather, Tijana Rajh, and Nada M. Dimitrijevic Iron(III)-oxo Centers on TiO2 for Visible-Light Photocatalysis. Chem. Mater. 2010, 22, 409–413.

64. Bingan Lu, Xiaodong Li, Taihong Wang, Erqing Xie and Zhi Xu. WO_3 nanoparticles decorated on both sidewalls of highly porous TiO2 nanotubes to improve UV and visible-light photocatalysis. J. Mater. Chem. A, 2013, 1, 3900–3906.

65. Papp, J.; Soled, S.; Dwight, K.; Wold, A. Surface acidity and photocatalytic activity of TiO2, WO3/TiO2, and MoO3/TiO2 photocatalysts. Chem. Mater. 1994, 6, 496−500.

66. Dang Sheng Su, Siglinda Perathoner, Gabriele Centi. Nanocarbons for the Development of Advanced Catalysts. Chem. Rev. 2013, 113, 5782−5816.

67. Alex T. Kuvarega • Rui W. M. Krause, Bhekie B. Mamba. Multiwalled carbon nanotubes decorated with nitrogen, palladium co-doped TiO2 (MWCNT/N, Pd co-doped TiO2) for visible light photocatalytic degradation of Eosin Yellow in water. J Nanopart Res (2012) 14:776–792.

68. Silvana Da Dalt, Annelise Kopp Alves and C. P. Bergmann CNTs/TiO_2 Composites. C. Avellaneda (ed.), NanoCarbon 2011, Carbon Nanostructures, DOI: 10.1007/978-3-642-31960-0_6, Springer-Verlag Berlin Heidelberg 2013.

69. Wang H, Wang HL, Jiang WF (2009) Solar photocatalytic degradation of 2,6-dinitro-p-cresol (DNPC) using multi–walled carbon nanotubes (MWCNTs)–TiO2 composite photocatalysts. Chemosphere 75:1011–1105.

70. N.R. Khalid, E. Ahmed, Zhanglian Hong, L. Sana, M. Ahmed Enhanced photocatalytic activity of grapheneeTiO2 composite under visible light irradiation. Current Applied Physics 13 (2013) 659–663.

71. Yongye Liang, Hailiang Wang§, Hernan Sanchez Casalongue, Zhuo Chen, and Hongjie Da. TiO2 Nanocrystals Grown on Graphene as Advanced Photocatalytic Hybrid Materials. Nano Res. 2010, 3(10): 701–705.

72. P. Muthirulann, C. Nirmala Devi, M. Meenakshi Sundaram. TiO2 wrapped graphene as a high performance photocatalyst for acid orange 7 dye degradation under solar/UV light irradiations. Ceramics International 40 (2014) 5945–5957.

73. Kefayat Ullah, Ze-Da Meng, Shu Ye, Lei Zhu, Won-Chun Oh. Synthesis and characterization of novel PbS–graphene/TiO2 composite with enhanced photocatalytic activity. Journal of Industrial and Engineering Chemistry 20 (2014) 1035–1042.

74. Yan SC, Li ZS, Zou ZG Photodegradation performance of g-C3N4 fabricated by directly heating melamine. Langmuir 25: (2009) 10397–10401.
75. Dong F, Wu LW, Sun YJ, Fu M, Wu ZB, Lee SC (2011) Efficient synthesis of polymeric g-C3N4 layered materials as novel efficient visible light driven photocatalysts. J Mater Chem 21:15171–15174.
76. Honglei Zhu, Daimei Chen, Du Yue, Zhihong Wang, Hao Ding. In-situ synthesis of g-C3N4-P25 TiO2 composite with enhanced visible light photoactivity. J Nanopart Res (2014) 16:2632–2642.
77. M. Radoicic, Z. Saponji, I.A. Jankovic, G. Ciri c-Marjanovic, S.P. Ahrenkiel, M.I. Comor Improvements to the photocatalytic efficiency of polyaniline modified TiO2 nanoparticles. Applied Catalysis B: Environmental 136–137 (2013) 133–139.
78. Seema Singh, Hari Mahalingam, Pramod Kumar Singh. Polymer-supported titanium dioxide photocatalysts for environmental remediation: A review. Applied Catalysis A: General 462– 463 (2013) 178–195.
79. S. Xu, Y, Zhu, L. Jiang, Y. Dan. Visible Light Induced Photocatalytic Degradation of Methyl Orange by Polythiophene/TiO2 Composite Particles Water Air Soil Pollut (2010) 213:151–159.
80. Y.F. Zhu, Y. Dan, Photocatalytic activity of poly(3-hexylthiophene)/titanium dioxide composites for degrading methyl orange, Sol. Energ. Mat. Sol. C 94 (2010) 1658–1664,.
81. X. Li, G. Jiang, G. He, W. Zheng, Y. Tan, W. Xiao, Preparation of porous PPy-TiO2 composites: improved visible light photoactivity and the mechanism, Chem. Eng. J. 236 (2014) 480–489.
82. Ariadne H. P. de Oliveira, Helinando P. de Oliveira. Optimization of photocatalytic activity of PPy/TiO2 nanocomposites. Polym. Bull. (2013) 70:579–591,.
83. Jianhong Wei, Qi Zhang, Yang Liu, Rui Xiong, Chunxu Pan, Jing Shi. Synthesis and photocatalytic activity of polyaniline–TiO2 composites with bionic nanopapilla structure. J Nanopart Res (2011) 13:3157–3165,.
84. M.A. Salem, A.F. Al-Ghonemiy, A.B. Zaki, Photocatalytic degradation of Allura red and Quinoline yellow with polyaniline/TiO2 nanocomposite, Appl. Catal. B Environ. 91 (2009) 59–66.
85. H. Zhang, R. Zong, J. Zhao, Y. Zhu, Dramatic visible photocatalytic degradation performances due to synergetic effect of TiO2 with PANI, Environ. Sci. Technol. 42 (2008) 3803–3807.
86. Rui Fang, Ying Liang, Xueping Ge, Ming Du, Shubiao Li, Tianyu Li, Zhi Li. Preparation and photocatalytic degradation activity of TiO2/rGO/polymer composites. Colloid Polym Sci (2015) 293:1151–1157.

GPSR Compliance

The European Union's (EU) General Product Safety Regulation (GPSR) is a set of rules that requires consumer products to be safe and our obligations to ensure this.

If you have any concerns about our products, you can contact us on ProductSafety@springernature.com

In case Publisher is established outside the EU, the EU authorized representative is:

Springer Nature Customer Service Center GmbH
Europaplatz 3
69115 Heidelberg, Germany

Batch number: 09484955

Printed by Printforce, the Netherlands